IMAGES OF ENGLAND

THE KING'S SHROPSHIRE LIGHT INFANTRY 1881–1968

The 1st KSLI in India in 1930.

IMAGES OF ENGLAND

THE KING'S SHROPSHIRE LIGHT INFANTRY 1881–1968

PETER DUCKERS FOR THE
SHROPSHIRE REGIMENTAL MUSEUM

First published in 1998 by Tempus Publsihing

Reprinted in 2009 by
The History Press
The Mill, Brimscombe Port,
Stroud, Gloucestershire, GL5 2QG
www.thehistorypress.co.uk

Reprinted 2010, 2012

ISBN 978 0 7524 1193 4

Typesetting and origination by
Tempus Publishing Limited
Printed in Great Britain.

The 1st KSLI 'forming square' in Malta in 1883.

Contents

The KSLI on Public Duties, led by two Welsh Guardsmen, at Buckingham Palace in 1948.

Introduction

The King's Shropshire Light Infantry (KSLI) existed as such for less than a hundred years – from 1881 until 1968. However, its origins go back well over a hundred years before that, to the middle of the eighteenth century. Its two constituent units, the 53rd (Shropshire) Regiment and the 85th (King's) Light Infantry, were raised in 1755 and 1759 respectively for service against the French in the Seven Years' War, 1756–63, and both campaigned in Europe, the West Indies and in North America prior to 1800.

The two Regiments saw a great deal of active service in the period 1793–1815. Both served in the Flanders campaign, 1793–5, before the 53rd sailed for India where it fought the Gurkhas of Nepal, 1814–17. A war-raised 2nd Battalion of the 53rd served through the Peninsular War from 1809 to 1814 and then left for St Helena in 1815 as Napoleon's guard. The 85th served in Walcheren, in Spain and in Southern France between 1809 and 1814 before leaving to join the British army fighting the United States. The Regiment fought at Bladensburg in 1814 and took part in the entry into Washington. Moving to the southern United States, the 85th fought in the disastrous battle of New Orleans in January 1815. It returned home to begin a very long period of peaceful garrison duty in Great Britain and throughout the Empire and was not called upon to serve on campaign until 1879, when it took part in the Second Afghan War.

The 53rd, on the other hand, spent a long period in India and saw a great deal of active service – in the two Sikh Wars, 1846–6 and 1848–9, on the North West Frontier, 1851–2, and through the arduous Indian Mutiny campaign, 1857–9. Here, it played a leading part in the operations around Lucknow in 1857 and 1858 and won five Victoria Crosses.

When the Regiments were chosen for amalgamation in 1881, the 53rd was in England and the 85th in South Africa. With the old 53rd forming the 1st Battalion and the 85th the 2nd Battalion, the new King's (Shropshire Light Infantry) moved into specially built barracks at Copthorne, Shrewsbury. It is at this point chronologically that the present selection of photographs begins.

With only a few exceptions identified in the text, the majority of the photographs reproduced here are taken from the archives housed in the Shropshire Regimental Museum in Shrewsbury Castle; many have never before been published.

Although photography has its origins in the 1840s, the KSLI collection has very little reflection of those days. The Regiment's photographic archive does have a few early examples though, its earliest photograph being a portrait of a recruiting Sergeant, *c.* 1850. But thereafter, photographic evidence of the activities and personnel of the 53rd or the 85th is scant until well into the 1870s. Indeed, it is not until the 1880s – when those two Regiments had been brought together to form the KSLI – that a more widespread use of photography begins to become apparent in the regimental archives.

Even so, it will come as no surprise that the archives are particularly lacking in some aspects; good quality photographs of the period 1880–1900, for example, are confined largely to studio portraits of officers or a few group photographs taken for the Regiment. For obvious reasons, there are few 'action' shots, taken on active service in, for example, the Boer War or in the conflicts of 1914–18 and 1939–45.

On the other hand, there are some areas where there is almost an embarrassment of riches – the collection has a comparatively large number of photographs of the North West Frontier in the 1930s and of India generally and becomes more comprehensive as it draws chronologically nearer to the present day. For accidental or historical reasons, some Battalions of the KSLI are more fully represented in the museum's collection and therefore in this selection. For example, there is hardly any material on the 2nd Battalion or the 8th in Salonika between 1915 and 1918, while there are many more photographs of the 7th KSLI during the First World War than of any other Battalion. Less surprisingly, the 4th Battalion is fully recorded from a photographic point of view; as a Territorial Battalion, it was made up of local men, many of whom have given their photographs or albums to the regimental archives over the years.

Using the collection as it stands, it would have been comparatively easy to present pages of group photographs of officers and men of the KSLI at various times and locations, or quantities of portraits and postcards of, usually unidentified, individuals in uniform. There might have been some virtue in this arrangement from the point of view of the family historian or collector trying to identify a particular person. But in setting out the selection of photographs offered here, I have tried as far as possible to avoid large numbers of simple portraits, whether of individuals or groups, unless they serve to illustrate a type of uniform or illuminate a particular occasion or event. Instead, an attempt has been made – given the limitations of the photographic collection now available – to show the officers and soldiers of the KSLI in a wide variety of roles and situations, in peacetime and at war, reflecting the range of their activities, experiences and achievements.

I hope that the photographs presented here will be of value to all those interested in the history and services of the King's Shropshire Light Infantry, whether they are former members of the Regiment, genealogists, military historians or the public at large.

Peter Duckers
Shrewsbury 1998

Acknowledgements

The compiler would like to thank the Trustees of the Shropshire Regimental Museum for permission to publish the photographs preserved in the museum archives and also the Imperial War Museum for their permission to reproduce those photographs indicated (IWM) in the text.

One

1881–1914

The Colours of the 53rd in 1881, the year of their amalgamation with the 85th to form the King's Shropshire Light Infantry. These Colours, presented in 1877, remained in service until 1954. On the back row, from left to right: Col. Sgt A. Brown, Col. Sgt Fortnum, Col. Sgt Davies. On the front row: Sgt Maj. Shortt, 2 Lt Wallace, Col. Sgt Gough, 2 Lt Pearce, Drum Maj. Tutton.

Officers of the 85th at Pinetown, Natal, in 1881. After serving in the Afghan War, 1879–80, the 85th was sent to South Africa, although it did not see any service in the disastrous campaign waged against the Boers of the Transvaal. The 85th returned to England in 1881 to amalgamate with the 53rd.

Officers of the new 1st Battalion of the KSLI in Malta in 1884. The officers wear blue patrol jackets with forage caps, or scarlet serge tunics with the white foreign-service helmet. Some wear the 1882 Egypt Medal and Khedive's Star.

The 1st Battalion at Fort Ricasoli, Malta, in 1883. During the War in Egypt in 1882, the 1st KSLI had served as part of the Alexandria garrison and in the occupation of Cairo. It received the first battle honour of the new Regiment – 'Egypt 1882' – as a result. The Battalion was stationed in Malta prior to its departure in 1885 for active service at Suakin on the Red Sea coast of the Sudan. The men are assembled to receive the Bronze Star which was awarded by the Khedive of Egypt for service in the 1882 campaign. The presentation was made by Major General the Hon. P. Fielding, commander of British forces in Malta.

Volunteers in the Hong Kong Plague. In 1894, during the 1st Battalion's tour of duty in Hong Kong, an epidemic of bubonic plague – the notorious 'Black Death' of the Middle Ages – swept through the colony. Nearly 3,000 Chinese citizens died of the disease before it was finally brought under control. Along with other volunteers, 300 men of the 1st KSLI took part in work to combat the plague, bury the dead and clean the streets.

In these photographs, volunteers of the KSLI are shown at work searching houses, clearing rubbish, cleansing streets and buildings and removing the dead – a very unpleasant business. Infected houses were disinfected with a mixture of chloride of lime and sulphuric acid before being boarded up. The men carried lamps to help find their way in the dark houses and streets. Special plague medals – gold for officers and silver for other ranks – were presented to the volunteers by a grateful Hong Kong community.

Officers of the 1st KSLI in Hong Kong, *c.* 1894. They display a mixture of clothing – civilian dress, scarlet tunics or blue patrol jackets and most wear forage caps. Seated in the front row, third from the left, is Capt. C.G. Vesey, the only officer of the battalion to die of bubonic plague in Hong Kong. One soldier of the KSLI also died of the disease during the epidemic.

Soldiers of the 2nd KSLI in action at Paardeberg in February 1900. This sketch by Harry Payne was produced specially for the Battalion which served throughout the Boer War from December 1899 to its conclusion in May 1902. British infantry wore a full khaki service dress by this date with a white cork helmet, camouflaged with a khaki cover.

'You have made a gallant defence, sir'. Lord Roberts of Kandahar received the surrender of the Boer Commander-in-Chief, General Piet Cronje, at Paardeberg on 27 February 1900. The 2nd KSLI served with distinction in the operations of 10 – 27 February which led to the surrender of the entire army under Cronje. This was the Boer army which had been besieging Kimberley. 'Paardeberg' became a regimental battle honour and 'Paardeberg Day' was celebrated annually.

The Volunteers in action in the Boer War. In this naive but spirited painting, officers and men of the 1st Volunteer Service Company are seen attacking a defended Boer farm. This War was the first occasion in which soldiers of the 1st and 2nd Volunteer Battalions of the KSLI – normally reserved for home defence – saw active service overseas. Two Volunteer Service Companies were formed, each serving for a year in South Africa. They earned the Volunteers' first battle honour, 'South Africa 1900–02'.

Officers and senior non-commissioned officers of the 2nd KSLI at Middelburg in the Transvaal in 1902. The Battalion played a full part in the wearisome 'guerrilla war' phase of the campaign, 1900–02, chasing fast-moving and elusive Boer columns across the vast expanses of the veldt.

Men of the KSLI in the 1st Mounted Infantry at Vereeniging in June 1902. The peace treaty was signed here in May. The large-scale use of specially trained Mounted Infantry was a key feature of the Boer War and a response to the mobility of the enemy. Their Commanding Officer, Capt. C.W. Battye, who was to receive the DSO for the campaign, is seated in the centre, with arms folded.

The 1st Volunteer Company at Shrewsbury station in 1901. A ceremony was held to welcome the men home after a year in South Africa. They wear their khaki drill uniforms and foreign-service helmets, as on campaign. The Guard of Honour, to the right, is made up of men of the 2nd Volunteer Battalion in their distinctive grey uniform.

Men of the 2nd Volunteer Service Company being welcomed home by the Mayor of Shrewsbury in 1902. The Mayor addressed the company from the Guildhall on The Square. These soldiers, who served for a year in 1901–02, wear the slouch hat adopted later in the war in place of the cork helmet.

The South African War Memorial on St Chad's Terrace, Shrewsbury. This fine representation of a KSLI soldier of the Boer War was formally inaugurated in July 1904 and records the names of all the officers and men of the Shropshire units who were killed or died in South Africa.

The Mounted Infantry of the 2nd Battalion, *c*. 1906. Mounted Infantry had been used on a fairly *ad hoc* basis by the British army since the campaigns in Afghanistan, 1878–80, and in Egypt, the Sudan and Burma in the 1880s. However, the experience of the Boer War led to a much more organized approach to their formation and training within each infantry battalion.

The 2nd KSLI on the parade ground in Fyzabad, India, *c*. 1906. The Battalion left South Africa in January 1903 to begin a tour of duty in India which was to last until 1914. The 2nd Battalion's strength was made up by drafts from England to number thirty-four officers and 1,113 men by March 1903, when it was stationed in Bareilly. Based in Fyzabad in 1905–09 and at Dum Dum and Dinapore in 1909–10, the Battalion was in Secunderabd when war broke out and it was ordered to mobilize.

The senior NCOs of the 2nd KSLI with Lt Col. J.L. Pearse, at the front, in the centre, *c.* 1908. Lt Col. Pearse was an officer of the old 53rd and 1st Battalion – hence he is wearing the Egypt/Suakin medals. He was appointed to command the 2nd Battalion in Fyzabad in February 1906 and remained in command until February 1910. The unnamed NCOs in this photograph wear the Queen's and King's medals for South Africa, some with the Long Service and Good Conduct medal. The Sergeant on the back row, far left, also wears the Distinguished Conduct Medal.

A mixed group, mainly from the 2nd KSLI, wearing badges of the Independent Order of Rechabites, 'Salford Unity', in India. Some are identifiable as 'Chief Ruler', 'Past Chief Ruler' and the like. The Rechabites was one of many temperance lodges established within the British Army in an attempt to counteract the incidence and effects of alcohol abuse.

A group of the 1st Battalion at Pembroke Dock in 1905. The 1st KSLI was stationed in the UK between 1903 and 1914 and took up residence at Pembroke Dock from 1903 to 1905. The men wear the peakless and very unpopular Brodrick cap which was introduced in 1902 and phased out in 1905.

SCOUTS 1ˢᵗ K.S.L.I. 1909

'Cads on Castors': cyclists of the 1st Battalion in camp at Church Stretton in 1909. Bicycles made their appearance in the army from 1885 onwards, especially in Volunteer units. From 1908, regular infantry battalions raised companies of cyclists, largely to serve as dispatch riders, messengers and, as here, battalion scouts.

Capt. R.P. Miles, 2nd KSLI, in 1912. In addition to his medals for South Africa, Miles wears the Delhi Durbar medal for King George V's coronation ceremonies at Delhi in 1911. He was one of only five soldiers of the KSLI to receive the medal as Superintendent of Gymnasia in Secunderabad. Miles died of wounds received in action in 1914.

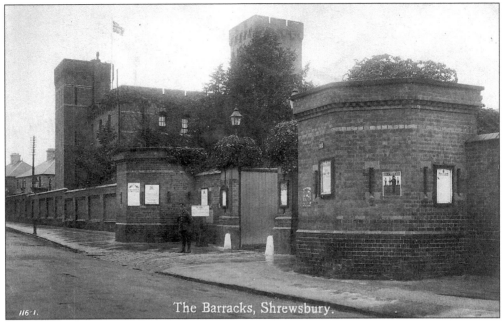

The main gate and guard rooms at Copthorne Barracks, *c.* 1912. New barracks were built at Copthorne between 1877–9 to accommodate the 53rd Brigade which included the newly amalgamated 53rd and 85th Regiments forming the KSLI.

A barrack room in the depot at Copthorne. This light and well organized room, with its folded beds and spacious look, reflects a great improvement on the old uncomfortable barrack blocks of earlier years.

24

The Mess Room, Shrewsbury Barracks.

A mess room in the depot, Copthorne, *c*. 1912. This is another reflection of the trend to pay more attention to the comfort and well being of the soldier which began with the army reforms of the 1870s.

The Sergeants' Mess, Shrewsbury Barracks.

The Sergeants' Mess, Copthorne Barracks. A comfortable room for relaxation, lined with prints, portraits and trophies.

The machine-gun platoon of the 2nd KSLI in India before the First World War. Note the pack-mules used to carry ammunition and the guns themselves.

A rifle-shooting team of the 2nd KSLI in India in 1911. Most of the men wear the distinctive Army Scout badge on the right sleeve. It was introduced in 1905 and designed by Robert Baden-Powell, the founder of the Boy Scout movement. The badge represented the 'North' pointer on a compass and was abolished in 1921.

A very fine study of a machine-gun crew of the 1st KSLI. The Maxim, seen here on its wheeled and armoured carriage, was introduced into general service with the British army in 1890; it was superseded by the Vickers in 1912. With a cyclic rate of fire of 500 rounds per minute, the Maxim played a major role in Britain's tropical campaigns in the late nineteenth and early twentieth centuries: 'Whatever happens we have got the Maxim gun and they have not'.

The 1st KSLI at Chelsea Barracks in 1909. From Hong Kong, the Battalion moved to India in 1895 and remained there throughout the period of the Boer War. In 1903 it returned to the UK to train and recruit and was stationed at Pembroke Dock, 1903–05, Bordon, 1905–07, and Lichfield, 1907–10. It spent one month, August 1910, in London where it relieved the Guards in Chelsea Barracks and undertook Public Duties at Buckingham Palace, St James's Palace and Hyde Park Magazine. Moving to Ireland in 1910, the 1st served in Fermoy and in Tipperary where it was based on the outbreak of war in 1914.

The soccer team of 'B' Company, 1st Battalion, in Fermoy in 1912. Sports of all kinds, especially rugby, hockey, cricket, football and boxing naturally played a large part in the life of the Regiment. Both Battalions had, and continued to have, a very fine record in most fields of sport.

The 1st KSLI near Farnham Park, *en route* to Aldershot, after serving as part of a Guard of Honour during the visit of the Japanese Prince Fushimi. The men wear the khaki service dress introduced in 1902 and look much the same as they did a few years later when marching off to war.

The 2nd KSLI at Secunderabad in 1914. Secunderabad was the largest military station on which the 2nd Battalion served during its tour in India and it was based there from 1911 until 1914. The city was home to a full division with two brigades of infantry, garrison forces, a brigade of cavalry and ancillary units. Here, in its last major piece of ceremonial duty before the War, and resplendent in white hot-weather uniforms, the Battalion marches past during the Queen's Birthday Parade. King George V and Queen Mary had actually visited India in 1911 for the Durbar festivities; they were the first reigning monarchs to do so.

Two

1914–1918

The **Shropshires** are with the rest
For King and Country they'll do their best.

Doing your bit: a patriotic postcard of 1914. It extols Shropshire's part in the 'greater conflict' and displays the flags of the nations allied against the Central Powers. Many varieties of such postcards were produced and were easily adapted to refer to any county or regiment.

Full dress in 1914. Although scarlet had been replaced by khaki for service uniforms, it was retained for full dress. Shown here is a bugler, a Private and an officer of the KSLI at the time of the outbreak of the War. Generally speaking full dress did not survive the First World War and scarlet was worn infrequently thereafter, other than by the band.

In service dress, 1914–16. A practical khaki uniform had replaced scarlet by the mid-1880s and had become standard by the time of the Boer War. The officer on the left is wearing his rank badges on his cuff. The Private in the centre is wearing the standard 1908-pattern webbing equipment which replaced earlier leather types. The Private on the right is wearing the Brodie steel helmet which the 1st KSLI helped to test under war conditions in 1915 before it was brought into general service in the winter of 1915–16.

A young soldier of the KSLI in 1914. This leather equipment was temporarily used in 1914, but not universally adopted, when supplies of the 1908-pattern webbing could not meet wartime demand. It proved unsuitable, however, for the wet and mud of the Western Front. The soldier is armed with the short magazine Lee Enfield .303 and the 1907-pattern 17 inch sword bayonet.

OUR TERRITORIALS
Don't beAlarmed
WE'RE
"On Guard"
at
SHREWSBURY

The Territorials, who were originally intended only for home service, are depicted here 'on guard at Shrewsbury'. In the event, Territorial Battalions served overseas from the beginning of the War and the 4th Battalion of the KSLI had a varied and distinguished career in the Far East and on the Western Front.

The Royal Visit, 1914, with King George V and the Mayor in The Square, Shrewsbury. The Guard of Honour in the background was formed by men of the 3rd Battalion, known as the Special Reserve since 1908, and was descended from the old county Militia. The 3rd KSLI spent most of the War in Ireland and was absorbed into the 2nd Battalion in 1919.

Soldiers of the KSLI in open countryside, before trench warfare, in 1914. The desolation of large areas of Flanders and France and the stagnation of trench warfare soon made such scenes little more than a memory.

Officers, NCO's and men of the 4th (Territorial) Battalion, with Indian servants, in Rangoon in 1914. The 4th Battalion had an exotic career after it left Shrewsbury for India in October 1914. It saw service in Rangoon, Hong Kong and Singapore, where it helped to suppress the Mutiny of 1915. Some of its men even got as far as Australia, escorting German prisoners from the cruiser *Emden*. This imperial garrison duty was a far cry, however, from the reality of warfare in Europe – as the 4th was to discover when it finally reached the Western Front in 1917.

Unknown soldiers of the war-raised Service Battalions of the KSLI. These men are typical of the hundreds of thousands who flocked to join their local regiments during the War.

Another example of a cheaply made, mass-produced and easily adapted postcard of 'the Great War'. Any regiment could be featured by changing only one line.

Don't be Alarmed the
Shropshire Light Infantry
are on guard.

Farewell to Bournemouth. Soldiers of the war-raised 7th Service Battalion cheerfully march to Romsey and Aldershot at the end of April 1915. They were to land at Boulogne in September as part of the 76th Brigade of the 25th Division. The 7th spent the entire War on the Western Front.

'In actual war': officers of the 2nd KSLI at the Front on Christmas Day in 1914, 'a few days after landing in France'. Note the wearing of sheep-skin *poshteens* (coats) and other cold weather dress. The 2nd Battalion landed at Le Havre in December 1914 and went to Belgium. Coming from the heat of India, it was ill-equipped for winter on the Western Front. Of the nineteen officers shown here no less than nine were killed in the War and that in a battalion which only served on the Western Front for a few months before moving to the 'quieter' theatre of Salonika.

All over by Christmas. A card sent home by Pte William Rogers which captures the early enthusiasm of the 'race for Berlin'. The 6th Division, in which the 1st KSLI served, is shown as the driving force of 'the Berlin Express'.

Regimental Sgt Maj. S.G. Moore, of the 1st Battalion, pictured just after being awarded the Military Cross for gallantry at Armentières in November 1914. This was a very early award of the newly instituted Military Cross and the first to be given to an RSM. The MC was infrequently awarded to senior NCOs and was usually given only to commissioned officers.

Officers of the 1st KSLI in France in 1916. The Battalion moved from Tipperary to Cambridge in August 1914 to join the 16th Brigade of the 6th Division. It landed at St Nazaire in September and spent the rest of the War in Flanders or France. The 1st served through the terrible battles in the Ypres Salient in 1915 and into 1916. This photograph was probably taken near St Omer in July 1916 when the Battalion moved to the Somme.

Brigadier General E.A. Wood CMG, DSO, Croix-de-Guerre. A former cavalryman, known as 'long pole' from his habit of carrying a cut-down lance in action, Wood commanded the 6th KSLI from 1915 to 1917 before succeeding to the command of the 55th Brigade. Twice recommended for the Victoria Cross, he was one of only seven officers serving in the First World War to receive the DSO and three bars.

The Welsby family of Shrewsbury exemplify patriotic fervour in action. All nine sons (one of whom is not present here) served in the army during the War, though not all in the KSLI. There are several other known examples of Shropshire families with six, seven or even eight members serving at one time. Mrs Welsby received a personal message from the King congratulating the family on their patriotic spirit and sense of duty. Remarkably, all nine survived the War which was not the case with many other local families.

Soldiers of the 4th KSLI relaxing at Tanglin Barracks, Singapore. Amateur dramatics had always been popular in the army, especially as a way of entertaining the troops when they were based in remote or fairly quiet garrison areas.

Men of the 4th KSLI at Katoomba, Australia. To soldiers of the KSLI serving in Europe, the movements of the 4th must have been followed with envy; the Far East was indeed far from the War. The 4th did, however, see some action in helping to suppress the mutiny of an Indian Regiment in Singapore in April 1915. The men pictured here served as an escort to German prisoners of war from the cruiser *Emden*, which was sunk in action by the Australian cruiser *Sydney* off the Cocos Islands. Two contingents of the 4th KSLI were the first British troops to serve in Australia since the formation of the Commonwealth of Australia in 1901 and they were given a tremendous reception.

A Christmas card produced by officers of the 6th KSLI to aid the Widows and Orphans Fund in 1916. The 6th Battalion had a 'Pals' element, 'C' Company, the only KSLI Battalion to do so. At the time of this card, with its stark image of soldiers plodding along the skyline, the 6th was near Amiens.

The 1st Battalion leaving the trenches near Ginchy in 1916. The Battalion took part in the Battle of the Somme in September 1916 and was engaged in the successful attack on 'the Quadrilateral' on 18 September and at Morval. This is a typical Western Front scene showing heavily laden soldiers crossing a devastated landscape. Note the stretcher-bearer in the background.

An early war grave with the original plaque and cross for Sgt V. Malt MM, of the 7th KSLI, killed in action on the Bapaume–Arras railway on 21 August 1918. After the War it was the task of what eventually became the Commonwealth War Graves Commission to concentrate and relocate graves, producing the familiar military cemeteries – 'the cities of the dead' – which still exist today.

Soldiers of the 7th KSLI snatching a rest on the way back from the trenches, photographed near Toutencourt on 18 May 1917. The 7th had been engaged in the Battle of Arras in April and in the fighting on the River Scarpe in May. (IWM)

An evocative study of men of the 7th Battalion, tired, unshaven and caked in mud, 'just after they had left the trenches' near Arras. The 7th took part in the Battle of Arras, fighting at Chapel Hill and Tilloy in April 1917 and then in the Third Battle of the Scarpe at Bois des Vert. In May the 7th was withdrawn for rest and further training and then returned to the Ypres Salient in September. Sgt W.J. Lewis, on the front row, second from the left, was awarded the MM in March 1917, but was killed in action on 28 March 1918 near Arras during the great German Spring Offensive.

Some of 'A' Company, 4th KSLI. The reality of warfare on the Western Front came to the 4th in the winter of 1917 when it returned from the Far East and was pitched into the Third Battle of Ypres. The men shown here are 'a few of the boys that are left' after the 4th's famous attack on Bligny Hill in June 1918. The Battalion was awarded the French Croix-de-Guerre for its part in the battle.

Battles of the Lys in 1918. Men of the 1st Royal Scots Fusiliers watch the 7th KSLI marching to the line near Oblenghem on 11 April 1918. Here, the Battalion took over trenches east of the La Bassée canal and near Locon and was heavily attacked on 12 – 13 April. Having suffered heavy casualties, the 7th returned to billets in Oblenghem; however, they were back in the trenches by 21 April. (IWM)

Finish Johnny! A humorous card sent home by Pte B.W. Gregory of the 8th KSLI. This Battalion, like the 2nd, spent 1915–18 on the Salonika Front. The card celebrates the defeat of the Bulgarians and 'Johnny Turk' which was achieved by October 1918.

sted in 4th Batt.
.I., Sept. 17, 1912

d up for Active
e, Aug. 4th, 1914

ngland Oct. 25th
1914.

Arriving at Burmah
Dec. 11th, 1914.

Went to Singapore the
following Spring.

Arrived in France,
July 30th, 1917.

IN LOVING MEMORY OF

ERNEST EDWARD (ERN)

The beloved and elder son of John & Emma Tipton, "Glendale," Annscro

Who was killed in France by shrapnel, on Oct. 31st, 19

AGED 23 YEARS.

Oft times we think of you dear Ern,
'Tis sweet to breathe your name,
In life we loved you very much.
In death we do the same.
When last we saw your loving face,
You were so strong and brave,
Without a smile or shake of hand,
You left us for a better land.

One of the many different types of memorial card produced by or for grieving relatives during the War. This one commemorates Ernest Tipton of Annscroft. Having served with the 4th through its Far Eastern tour of duty, Tipton was killed at Passchendaele on 31 October 1917. This was the Battalion's first day of action on the Western Front.

Sgt Harold Whitfield receiving the Victoria Cross from the King in France in 1918. Whitfield, of the Shropshire Yeomanry and its dismounted successor, the 10th KSLI, was awarded the VC for his gallantry in the single-handed capture of a Turkish machine-gun position at Birj-el-Lisaneh, Palestine, on 10 March 1918. Whitfield was the only man in the Shropshire Regiments to be awarded the Victoria Cross in the First World War.

Three

1919–1939

Journey's End: a Christmas card for 1918. It was produced by the 2nd KSLI to record its movements from India in 1914, via Ypres, the Somme, Salonika and Shrewsbury to Cork. The 2nd remained in Ireland until December 1922 by which time, with the formation of the Irish Free State, all British troops had left.

Army of Occupation: the 1st KSLI being inspected by General Plumer at Imsdorf in 1919. The Battalion's last actions in the War were in the Battle of the Selle late in October 1918. It was at Bohain when the War ended and it joined the 2nd Army in the general advance into Germany, crossing the frontier on 17 December. The 1st then served briefly as part of the Rhineland occupation force before returning to England and to Oswestry in April 1919. Reduced to a mere cadre by the rapid process of post-war demobilization, the 1st KSLI was entirely re-formed in 1919.

Army of Occupation: the 2nd KSLI leaving Minden Barracks, Cologne, in 1926. The 2nd became part of the Rhineland occupation force in 1924 and served in Cologne until January 1926, when the Cologne Bridgehead was evacuated. The Battalion then moved to Wiesbaden.

Army of Occupation: field kitchens and horse transport of the 2nd KSLI outside Wiesbaden Barracks in the Rhineland in 1927. The Battalion returned to Aldershot in November of that year.

Ellesmere College Cadet Corps in the 1920s. Cadet Corps had been established on a regular basis since 1900, as part of the patriotic response to the Boer War; the corps at Ellesmere College was the first to be formed. Thereafter, most of the county's public and grammar schools raised their own Cadet Corps; they continue to exist to the present day.

A recruiting drive in Shrewsbury, with soldiers of the KSLI showing off a new Universal Carrier to an interested group of onlookers.

Lt Col. J.C. Hooper (centre) with the team of the 2nd KSLI which won the Army Hockey Championship 1926–7. This was an achievement which it attained on a number of other occasions. On the back row, from left to right: Pte Edwards, Lt Barlow, Sgt Jones, L Cpl Attwood, Bandsman Biddulph, Sgt Roberts. On the front row: CQMS Herbert, Lt Ely, Capt. Eager, Lt Col. J.C. Hooper, Lt Tarr, Sgt Bugler Perry, CQMS Davis.

Private Samuel Hartley, of the 1st KSLI, showing off his tropical or warm weather kit. This white uniform was worn on ceremonial occasions in the hot season in India. The 1st KSLI went from England to Aden in 1919 and from there to India where it served until 1938.

Buglers of the 2nd KSLI, *c.* 1933. The Battalion returned from Germany in 1927 and was based at Aldershot until 1931 when it moved to Sobraon Barracks, Colchester. Traditionally, bandsmen wore a more colourful and distinctive uniform than other soldiers, as witnessed by the striped shoulder wings visible here, and after the First World War they retained the scarlet dress which was more or less abandoned by the rest of the army. Seated fourth from the left is Lt Col. L.H. Torin MC, who assumed command of the 2nd Battalion in May 1933. He and the other officers are wearing the Home Service helmet which was first introduced in 1878.

The Band of the 1st KSLI in Rawalpindi, *c.* 1931. Apart from entertaining or marching with their own comrades, bands of British Regiments on service in India, and elsewhere throughout the Empire, frequently gave public concerts or played at social gatherings of all kinds. The two officers seated to left and right of the bass drum are, on the left: Lt Col. H.A.R. Aubrey OBE MC, who assumed command of the Battalion at Rawalpindi in January 1930, and on the right: Major G.S. Brunskill MC, who succeeded him in command of the 1st Battalion in 1934.

Drill Order in India in the 1930s. This Private wears his shirt open at the neck and folded in, with sleeves rolled up. The Wolseley-pattern tropical helmet is of the type which remained in use until 1942 when a lighter version was introduced. He wears shorts with puttees and hose-tops.

Another view of the same soldier, showing the 1908-pattern webbing equipment which was still in use. On either side of the belt there are five fifteen-round ammunition pouches, with a haversack hanging on the left and a water bottle on the right.

Men of the 1st Battalion resting near Tauda China in 1931. The Battalion was called upon to do several tours of duty on the North West Frontier from November 1929 onwards and served in the most hostile area – Mahsud territory in Waziristan, where this photograph was taken. The village of Tauda China was destroyed in a punitive raid.

Soldiers of the 1st KSLI with a Lewis gun in a *sangar* (a stone breastwork) on the perimeter defences of Saidgi camp, lower Tochi, during frontier operations in 1931. The Battalion received the India General Service Medal with clasp 'North West Frontier 1930–31' for its services in Waziristan, against the Afridis on the Khajauri Plain and in the 'Red Shirt' rebellion.

Razmak in the snow, 1930–1. The great frontier base at Razmak was built in the heart of tribal territory in Waziristan. It was begun in 1922 and designed to intimidate what had been the most persistently hostile of the frontier tribes. The garrison consisted of a brigade headquarters, a mountain-artillery brigade, six British and Indian infantry battalions and ancillary troops. This was a huge commitment of regular forces deep in 'enemy' territory. Sniping at the Razmak Movable Column became a popular pastime as it periodically marched through the district to show the flag and keep the routes open.

The 1st KSLI on the march near Tundan on the North West Frontier. On 'Road Open Days', columns were employed to march from base to base to make sure that routes were open and to remind local tribesmen of the presence of British and Indian forces in an attempt to keep them in check.

A column camp, which included the 1st KSLI, at Idak on the frontier in 1931. The fort dominating the heights is a typical feature of the frontier. These frontier forts were sometimes manned by regulars although the usual garrison was made up of frontier scouts, frontier constabulary or even *khassadars* (a form of tribal militia). They provided a network of points of communication, observation and control.

The Officers' Mess of the 1st Battalion: a Guest Night at Kamptee in 1935. The officers wore regimental mess kit and were attended by mess waiters and Indian *khidmatgars* (servants). Apart from its frontier service, the 1st KSLI carried out a regular round of garrison duties and training in India. It was stationed in Poona, Dinapore, Rawalpindi, Delhi, Kamptee and Karachi. As was usual, periods of hot weather were spent in cooler hill stations like Gharial and Chakrata and in the cool season there were exercises and manoeuvres. There was also time for traditional sports and more typically Indian activities like pig sticking, tiger hunting and polo. While the Battalion was stationed in Delhi, it was also called upon to carry out the usual ceremonial functions associated with the office of the Viceroy.

Abdul, a *khidmatgar* who served with the 1st Battalion throughout its service in India. He wears a smart starched white uniform, without shoes, and on his *pagri* (head dress) he wears a special silver version of the regimental badge. Note that he wears the campaign medal for 1930–1. As a non-combatant 'follower' he was entitled to the medal if he served with the Battalion in the designated campaign area.

Officers and NCOs of 'A' Company, 1st Battalion, in 1931. This picture was taken at Karawal Camp on the Khajauri Plain, south west of Peshawar, during the 1930–1 Afridi operations. The men wear the button-up jersey adopted for the cold season in India.

The Indian Platoon of the 1st KSLI at Karachi in 1938. Every British infantry battalion stationed in India after the First World War had an Indian Platoon serving with it. The 1st Battalion's Indian Platoon was established in 1922 and remained with the Battalion until the army reorganization in 1938, when the 1st was in Karachi. The reorganization led to the disbandment of the Indian Platoons and the men returned to their former Indian Army units. Seated on the front row, fourth from the right, is Naubat Singh, the platoon's Subadar Major (the senior Indian officer) and 'a great friend of all'.

Public Duties at the Viceregal Palace, Delhi, in 1934. These men of the Band and Bugles wear the white ceremonial uniform worn during the hot season. The 1st Battalion was in Delhi for eighteen months during 1934–5 and suffered a constant demand for men to perform ceremonial duties. This caused great problems with the Battalion's training programme.

Training with poison gas: the decontamination squad of the 1st Battalion at Belgaum in 1935. The threat of poison gas was taken very seriously after the First World War and even in India, anti-gas schools and training establishments were set up. There was a Chemical Defence Research Establishment at Rawalpindi and an Anti-Gas Training Wing at Belgaum.

The Regimental Police of the 1st Battalion at the hot-weather hill station of Gharial in 1931. They wear the distinctive 'RP' cuff band. On the back row, from left to right: Pte Edwards, Pte Davies, Pte Coulson, Pte Harman, Pte Whenman. On the front row: Pte Davies, Pte Embrey, Sgt Tomkins, L Cpl Morris, Pte Edwards.

Signallers of the 1st Battalion training in India in 1935. Even with the advent of portable radios, the semaphore system was much used on the frontier. The heliograph was more effective in the bright light of the tropics, however, and it carried much further – when the sun was shining.

The Band and Bugles of the 1st KSLI at Karachi in 1935. Karachi was the last posting of the 1st Battalion on the subcontinent. The old 53rd had seen extensive and arduous service in India in the periods 1805–23 and 1844–59 and the 85th had done tours of duty on the subcontinent in the 1860s and 70s. However, the Regiment never returned to India after the 1st Battalion's departure from Karachi in 1938.

The King and Queen visiting Bordon in 1939. The 1st Battalion returned from India on the *Neuralia*, the same ship which had brought the 2nd Battalion from India in 1914, and went to Bordon camp. The Battalion was reorganized and became part of the 3rd Infantry Brigade, 1st Division. While it was training at Bordon, the camp was visited by HRH King George VI and Queen Elizabeth, who is seen in this photograph receiving a bouquet. Lt Col. J.M.L. Grover, commanding the 1st Battalion, is standing in the background. While the Battalion was based here, the Second World War broke out and the Battalion was ordered to mobilize.

Four

1939–1945

Issuing kit in the quartermaster's store at Copthorne Barracks just before the War.

'D' Company of the 1st Battalion at Cysoing, France, in 1939. The 1st KSLI left for Cherbourg in September and joined the British Expeditionary Force. The men still wear the old service dress uniform with puttees. From its billets at Cysoing, the Battalion was one of the first British units to serve on the Maginot Line and arrived at Metz on 27 November. After the 'phoney war', during which time the Battalion underwent continual training, the rapid German advance and subsequent French collapse in May 1940 forced the retreat to Dunkirk. The 1st KSLI played a prominent part in the retreat, in the rearguard fighting and in the evacuation. It was one of the last British units to leave the beaches and harbour at Dunkirk in June 1940.

Cpl T.W. Priday, 1st KSLI, the first British casualty of the Second World War. Priday was killed by a mine while on patrol near Metz on 9 December 1939. He is buried in Luttange War Cemetery.

'A' Company, 1st KSLI. These soldiers had taken part in the campaign in France and the retreat to Dunkirk. The men were photographed after their return to England and are wearing the newly issued battle dress uniform with gaiters instead of puttees.

Amphibious training in Curaçao in 1941. The 2nd Battalion had a much more peaceful start to the War. From May 1940, it was based in the Dutch West Indies at Curaçao and Aruba, guarding oil refineries. The Battalion returned to England via New York in March 1942.

War-raised units of the KSLI. As in the First World War, new battalions of the KSLI were quickly formed. This photograph shows men of the 7th KSLI, raised in May 1940. The unit was converted to artillery, as the 99th Anti-tank Regiment, in 1942. It saw no overseas service, however, and was disbanded at Brigg, Lincolnshire, in December 1943.

Soldiers of the 5th KSLI watching a Bren gun demonstration by Pte Sam Corbett and admired by their mascot Billy the Bear. The 5th was formed in April 1939 as a duplicate of the 4th (Territorial) Battalion, but saw no active service and became a Home Defence and Training Battalion for the KSLI and the North Staffordshire Regiment.

Men of 'Y' Company, 6th KSLI, enjoying a cup of tea from a YMCA van at Anderby Creek on the East Coast in 1941. The 6th was raised in 1940 for Home Defence duties, but was converted to 181st Field Regiment, Royal Artillery, in March 1942. It went on to serve through the Normandy campaign in 1944 and in Belgium and Holland and was the first Field Regiment, Royal Artillery, to cross the Rhine. It was disbanded after the War in January 1946.

The 9th Salop (27th GPO) (Specialist) Battalion of the Shropshire Home Guard. This Specialist Battalion is shown here during exercises in The Quarry, Shrewsbury, in 1941. It was comparatively well equipped with regular-pattern battle-dress, boots, anklets, respirators and webbing equipment; other Shropshire Home Guard Battalions were not so fortunate in the early days. Note the distinguishing triangle on the sleeve of the man on the right and the 'SHR 9' district/unit badge. Nurses of the Voluntary Aid Detachment (VAD) are also in attendance.

An inspection of HQ, Shropshire Home Guard, in The Quarry, Shrewsbury, in 1941. The county raised eleven battalions of Home Guard and no less than 31,000 men went through its ranks in the period 1940–4.

High Street, Shrewsbury: a battalion of the Shropshire Home Guard marches past the Guildhall where the Mayor takes the salute. This was the 'stand down' parade of the Shropshire Home Guard in 1944.

Soldiers of the 1st KSLI talking to local Arabs beside a knocked-out German tank at Tebourba, Tunisia, in May 1943. After Dunkirk, the 1st Battalion was re-formed and re-trained over the course of nearly three years. In February 1943, under Lt Col. J.G. James, it sailed for North Africa to join the 1st Army and took part in the severe fighting in Tunisia. This led to the capture of Tunis and the expulsion of Axis forces. The 1st KSLI received battle-honours for Gueiriat el Atachi Ridge and Djebel Bou Aoukaz, for Tunis and the general theatre award, 'North Africa 1943'. (IWM)

A few souvenirs of North Africa: soldiers carrying back German rifles, helmets and gas-masks as trophies of war.

The 1st KSLI disembarking from landing craft on the island of Pantellaria, 11 June 1943. From North Africa, the Battalion took part in the opening moves of the campaign against the Italian mainland beginning with the seizure of the Mediterranean island of Pantellaria as a preliminary to the attack on Sicily. The 1st KSLI trained hard for this combined operation, but in the event the landing was unopposed and the island surrendered with only a few shots fired – and to the apparent delight of the inhabitants. (IWM)

Soldiers of the 1st Battalion take position beside a knocked-out coastal defence gun on Pantellaria. The island's defences had been heavily bombed from the air and shelled from the sea prior to the landing. The only opposition came from one pill-box which was silenced by men of 'B' Company. After three days on Pantellaria, the 1st KSLI moved back to North Africa for further training. (IWM)

Men of the 1st KSLI crossing a stream during fighting in the Anzio beach-head in March 1944. The battalion landed in Italy in December 1943 and was involved in the severe fighting following the landings at Anzio in January 1944. It served in the 'trench warfare' on the beach-head until the breakout was finally possible in May and the advance on Rome could continue. (IWM)

Taking a break: a group of 1st KSLI soldiers resting during the fighting in 1944. Standing at the back, on the left is Jack Massey; at the front on the left is Tommy Crick; at the back on the right is Pte A.G. Burt; at the back, second from the right is Pte Ianto Evans. Most of these men were from the Anti-tank Platoon.

Monte Battaglia, October 1944. The 1st KSLI took over the line from the Grenadier Guards. Rome fell to the Allies in June 1944, but the fighting in Italy did not end with the fall of the capital. The 1st Battalion served in operations against the Gothic Line and from August 1944 was involved in severe mountain fighting over the most difficult terrain. The Battalion did not witness the end of the War in Europe, however, as it sailed for Palestine in February 1945.

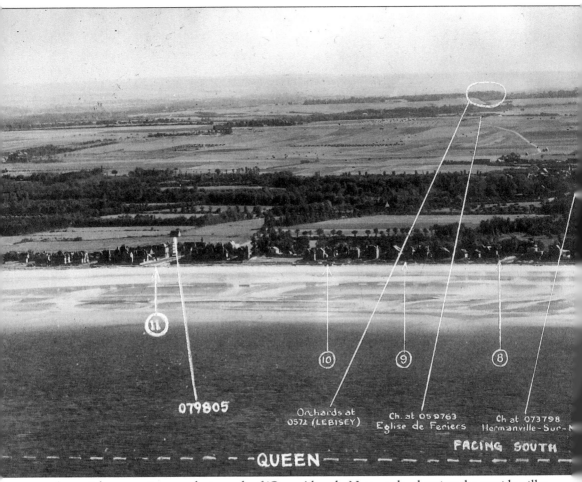

079805

⑪

⑩ Orchards at
    0572 (LEBISEY)

⑨ Ch. at 059763
    Eglise de Feriers

⑧ Ch. at 073798
    Hermanville-Sur-N

FACING SOUTH

- - - QUEEN - - - - - - -

An aerial reconnaissance photograph of 'Queen' beach, Normandy, showing the seaside villas of La Breche where the 2nd KSLI landed on D-Day, 6 June 1944. The Battalion had returned from the Dutch West Indies in 1942 and spent two years in rigorous training. After landing on 'Queen' beach, the 2nd established a firm base at Bieville and after a month of consolidation it took part in the break-out fighting at Caen and Manneville. By 1 September, the 2nd KSLI was at Villers-en-Vexin and was able to rest before the 'triumphal advance' through Belgium began.

German prisoners taken by the 2nd KSLI in the first day's fighting in Normandy. The Battalion reached Lebissey Wood on the outskirts of Caen on the first day which was the furthest inland point reached by any battalion on D-Day. (IWM)

The 4th KSLI at St Charles de Percy in August 1944. The 4th (Territorials) landed in Normandy on D-Day plus eight (14 June) and concentrated at Cainet. In this photograph German POW's are seen behind the anti-tank gun, guarded by Provost Sgt Bradbury (with the map case). Leaning on the gun barrel is Capt. 'Buster' Walford, second-in-command of 'A' Company, who was killed shortly after this photograph was taken. The tanks are those of the 3rd Royal Tank Regiment. (IWM)

Men of 'B' and 'D' Companies, 4th Battalion, supported by a section of carriers, moving through the deserted town of Vassy on 15 August 1944. The battalion dug in near the town prior to fighting its way across the River Vire.

Tired soldiers of 'B' Company, 4th Battalion, snatching some rest near Estrey. Note the sentry on watch, top left, and the 'Charging Bull' arm badge of the 11th Armoured Division.

Cheerful soldiers of the 4th Battalion crossing the Orne at Ecouché, south of the Falaise Pocket, prior to the attack on Montgaroult and Sentilly, 19 August 1944. The 4th then passed through Aubusson, Flers and St Honorine and went on to Laigle where it was able to rest and re-equip. (IWM)

First into Antwerp: men of Support Company, 4th KSLI, were welcomed in to the city on 4 September 1944. Antwerp was liberated by units of the 11th Armoured Division after a very rapid advance from the Seine which took them via Amiens into Belgium. The crew here includes Sgt Hayward, Pte Bassett, Pte Collis, Pte Green.

Sgt 'Jock' Wright, Battalion Armourer with the 4th KSLI, with the armourer's shop, checking Bren guns near Duerne in November 1944. (IWM)

Winter on the Maas: both the 2nd and the 4th KSLI spent the winter of 1944–5 facing the River Maas. This photograph shows soldiers of the 4th Battalion on the desolate west bank of the Maas near Broekhuizen. The advance began again in February 1945.

Viscount Gort inspecting recruits of 20th Infantry Training Centre (ITC). While the War went on overseas, the work of the Depot continued. There were major expansions to accommodate recruits. The Maltings in Ditherington, Shrewsbury, which was the world's first steel-framed building, was adapted to house the 20th ITC where recruits were trained for the KSLI and North Staffordshire Regiment.

A familiar sight at Ditherington and around Shrewsbury was 'Jessie' the donkey, seen here harnessed to a delivery cart, with Pte 'Hangman' Ellis. The donkey and cart were used to deliver the mail and to serve as an example of what could be done to economize on transport and fuel.

Field Marshal Montgomery decorating Sgt George Harold Eardley, 'A' Company, 4th KSLI, with the ribbon of the Victoria Cross. Born in Congleton in 1912, Eardley was transferred into the 4th KSLI in 1944 and had already won the Military Medal in August during the fighting at Beny-Bocage. On 16 October, he single-handedly attacked and destroyed three enemy machine-gun posts located in orchards near Smakt, east of Overloon. These posts were holding up the advance and successive efforts to destroy them had failed. He received the VC and the MM from the King in February 1945. Eardley worked as an electrical engineer for Rolls Royce after the War and died in Congleton in 1991.

Pte James Stokes, 2nd KSLI, was killed in the action for which he was awarded a posthumous Victoria Cross. After crossing the Maas at the end of February 1945, the 2nd advanced on Kervenheim. On two consecutive occasions on 1 March, with his platoon pinned down by heavy fire from fortified farm buildings, Stokes single-handedly dashed on ahead, cleared the buildings and brought back prisoners. Despite several severe wounds, he refused to go back for medical attention and in a third attack, shown here, he was mortally wounded. His men advanced past him and took the position. Stokes, a native of Glasgow, was buried in the Reichswald War Cemetery. This is a painting by Terence Cuneo commissioned by the KSLI.

Grand Admiral Karl Dönitz, General Jodl and Albert Speer were taken prisoner at Flensburg on the Danish border, 23 May 1945. Under Brigadier J.B. Churcher, the 159th Infantry Brigade, which included the 4th KSLI and Herefords, occupied the enclave which housed the seat of the remnant German government under Dönitz, based at Flensburg Castle. Dönitz had been Hitler's chosen successor as leader of the Reich, though not with the title 'Fuhrer'. In May 1945 in 'Operation Blackout', soldiers of the 159th arrested Dönitz, Jodl, Speer, Admiral von Friedeburg and the entire 'government'.

Five

1945–1968

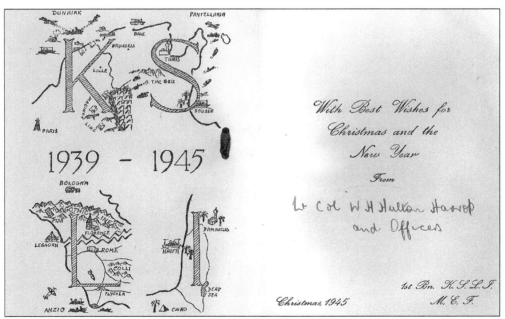

A regimental Christmas card from 1945 showing the campaigns of the 1st Battalion. From France in 1940, the Battalion campaigned through Tunisia to Pantellaria in 1943, Italy, 1943–5 and on to Palestine.

Other ranks of the 1st Battalion in Palestine in 1945. The Battalion moved from Italy to Nathanyu, Palestine, in February 1945 and became involved in the Arab-Jewish conflict. The 2nd KSLI also served in Palestine in 1945. In this photograph are, at the back, from left to right: CSM J. Baughan MM, Sgt C. Wilson MM, Pte J. Lowe. At the front: Sgt A.J. Speake, Lt Cpl Downes MM, RSM G. Gough MM, QMS D. Morgan BEM.

A Guard of Honour for Field Marshal Montgomery, saluting in the foreground, provided by 1st KSLI in Khartoum in December 1947. The 1st Battalion moved from Palestine to the Sudan in September 1946 and returned to England in 1948.

The Royal Hospital, Chelsea, 28 September 1947. The Governor of the hospital is seen with Major General J.M.L. Grover, Colonel of the Regiment, handing him the Colours of the Harford Light Dragoons and the Jamestown Light Infantry. These two US standards had been captured by the 85th at Bladensburg in 1814. For over a century they had hung in the Royal Hospital and are now in the Regimental Museum in Shrewsbury Castle.

The 1st Battalion returning home from Khartoum in 1948. In line with post-war army reductions, it was proposed that the 1st and 2nd Battalions should amalgamate in 1948. Accordingly, the 2nd KSLI returned home from Cyprus via Egypt in April 1947 and the 1st joined them in England in March 1948.

Long service and good conduct: Private 'Nick' Carter. Seen in this 1948 photograph are, from left to right: the Earl of Powis (Lord Lieutenant of Shropshire), Major General Grover, Pte Carter, Lt Col. A.S. Shaw-Ball. Pte Carter served in the 2nd Battalion and Depot from 1901 to 1951. He saw active service in the Boer War and in France, 1914–16, and thereafter served continuously with the 2nd Battalion until 1942, when he was posted to the Depot. He wears ten Good Conduct chevrons on his left sleeve indicating over forty years' service and at the time of his discharge, in 1951, he was the oldest serving soldier in the British Army. He remained, by choice, a Private throughout his career.

Amalgamation Day in 1948. Post-war reductions in the defence budget forced the amalgamation of many two-battalion Regiments, including the KSLI. The ceremony of amalgamation took place at the Depot, Copthorne, on 12 June 1948. The Colours of the two Battalions are seen here, paraded before the distinguished company present for the ceremony.

The Colours of the 1st and 2nd Battalions at the Amalgamation Parade. The Colours are, from left to right: 1st Battalion King's Colour (carried by Lt J.C. Evison), 1st Battalion Regimental Colour (carried by Lt B.V. Houghton-Berry). In the middle stands Major General Grover. Next to him are: 2nd Battalion King's Colour (carried by Lt D.K. Maclachan) and 2nd Battalion Regimental Colour (carried by Lt M. Cooper).

Public Duties in London in 1948–9. The first ceremonial role of the 1st KSLI – as the Regiment was to be called after the amalgamation – began in September 1948 when it was required to undertake Public Duties at the royal palaces – Buckingham Palace, St James's and Windsor Castle – and on Royal occasions in London. At this time the Brigade of Guards was overseas, campaigning in Malaya. The KSLI had last carried out such duties in 1909 (see p. 28) and displayed alternately, the 1st Battalion Colours (which had been used in 1909) and the old 2nd Battalion Colours. 'C' Company is seen here entering the forecourt of Buckingham Palace as the King's Guard. This was the first occasion on which the Light Infantry beret was worn by all ranks on Public Duties.

After its distinguished service in North West Europe, 1944-5, the 4th (Territorial) Battalion continued its peace-time activities and training. Lt Col. R.F. Arden-Close is seen standing in the centre, shortly after his appointment as Commanding Officer, with 'C' Company at Wellington in 1951.

Overseas again, this time in Hong Kong, 1949–51. Field Marshal Sir John Harding, Commander-in-Chief Far Eastern Land Forces, inspects a KSLI Guard of Honour in 1949. Note the Golden Cockerel badge of the 40th Division in which the KSLI was then serving.

A new Regimental Chapel. While the regulars were serving overseas, a new Regimental Chapel and memorial for the KSLI and Herefordshire Light Infantry was dedicated in St Chad's church, Shrewsbury. The 4th KSLI and the Herefordshire Light Infantry march past the Mayors of Shrewsbury and Hereford and other local dignitaries on Sunday 10 June 1951.

A view of the Regimental Chapel in St Chad's, sited in the existing Lady Chapel of the church, 10 June 1951.

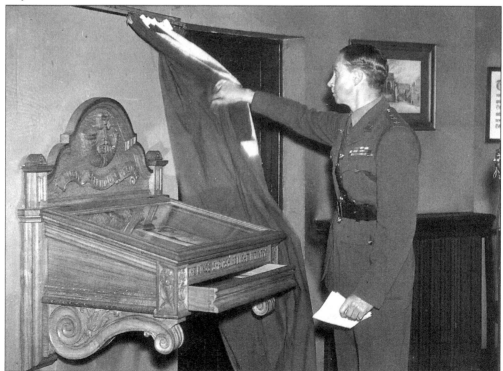

The memorial cabinet and Book of Remembrance at St Chad's. Servicemen from Shropshire have served in all parts of the world; some of those who never returned are commemorated in this Book of Remembrance.

The first UN war: Korea, 1951–2. The KSLI went from Hong Kong to the War in Korea in April 1951, serving in the 28th Commonwealth Infantry Brigade. The KSLI shares with the KOSB the record for the longest period of service in the Korean War. Shown here receiving 'specialist training' is a Private of the Royal Australian Regiment, his instructor being Cpl A. Tucker of the KSLI.

Sgt Custance 'and team' with a jeep-mounted Vickers' machine-gun in Korea. This machine-gun, whose first types were introduced in 1912, saw service through both World Wars and continued to be used until 1968. It had an *extreme* range of 4,500 yards.

The Lewendon brothers – imaginatively nicknamed 'Big Lew' and 'Little Lew' – with a Vickers' machine-gun in Korea. Each Battalion had six Vickers' machine-guns at the time of the Korean War. Note the 'parabellum' flash eliminator which had been in use since the end of the Second World War.

Taking cover: soldiers of the KSLI in a newly-captured trench enduring a Chinese mortar attack. This was during the first major action by the 1st Commonwealth Division. In October 1951, fighting was concentrated on hills known as Point 335, Point 208 and Point 227. The fighting here was the most severe the Battalion encountered in Korea. As one official account put it, 'heavy mortar fire greeted the forward British troops taking over the Communists' dug-outs'. Here, Pte D.B. Pikes of 'D' Company takes cover during one such barrage, with 'Bomber' Wells behind him.

On the watch in 'No-man's Land'. From left to right: Pte Denys Fowles, Lt N.C. Rowe, Pte Donald Freeman on reconnaissance.

A KSLI Bren gun position in Korea. The Bren gun had been introduced in 1938 as a light machine-gun and remained in use well into the 1960s. Its cyclic rate of fire was 450–550 rounds per minute.

The onset of winter at the HQ of the KSLI in Korea in November 1951. The weather was especially severe that winter; as the regimental history relates: 'it rained and snowed then thawed and snowed and froze again; everyone dug hard and the bunkers became deeper and deeper'.

A well protected dug-out in warmer days. The 'office' of the Signal Platoon in Korea. Korean porters and labourers are seen with men of the KSLI. At the back on the right, with a pipe, is 2nd Lt G.M. Benson and seated, at the front on the left, is the Signals' Officer, Capt. James Green.

Guarding prisoners of war. Men of 'B' Company embussed ready to move to their campsite on Koje-do Island in May 1952. They were to assist Canadian and US troops in guarding North Korean and Chinese prisoners held in the island camp.

A view of one of the prison compounds on Koje-do. The camp was described as 'a crazy island of dust, dirt and evil smells'. 'B' Company shared responsibility for Compound 66 which held 3,200 recalcitrant and disgruntled prisoners.

Coming home to Lichfield in November 1952. Sixty-two officers and men of 1st KSLI were killed or died in Korea. The Battalion left Pusan in September 1952 and reached Whittington Barracks, Lichfield, in November. On 6 November a representative detachment, shown here, left for a civic reception in Shrewsbury. It then went to Hereford on the 7 November for a similar series of functions.

A royal visit to Shrewsbury. Her Majesty the Queen and Prince Philip inspected a Guard of Honour of the 1/2 Cadet Battalion of the KSLI in Shrewsbury Castle on 24 October 1952. The 1/2nd was formed as an amalgamation of the 1st and 2nd Cadet Battalions in 1949. There were then twenty-three local or school companies of cadets throughout the county.

Göttingen on 15 October 1954. After six months in England, the KSLI left for Germany in April 1953 to join the British Army on the Rhine. They were stationed in the small market town of Göttingen, sixty miles south of Hanover and at the junction of the American, British and French zones of occupation. While here, the old Colours of the 1st and 2nd Battalions were trooped for the last time and new Colours were presented, 'a very great day in the annals of the Regiment'. The new Colours were presented by the Chief of the Imperial General Staff, Field Marshal Sir John Harding, before a distinguished gathering and over 300 guests.

The Bicentenary Parade in 1955. Having returned to England from Germany in March 1955 the Regiment celebrated its two 200th birthday, based on the raising of the old 53rd in 1755. The main event was the parade in The Quarry, in which the 4th Battalion, the Home Guard (which was re-raised in 1952) and Shrewsbury Cadets were also involved. Over 300 Old Comrades, pictured above, took part in the March Past and included no less than twelve former COs. The Regiment was honoured with the Freedom of the Borough and in a final procession the old Colours of the 53rd, presented in 1877 (see p. 9), were laid up in St Chad's church. Similar ceremonies were enacted at Bridgnorth, the birthplace of the 53rd, and in Hereford.

Off to Kenya in May 1955. The KSLI *en route* from Liverpool to Mombasa on the *Empire Halladale*. The Regiment, now 900 strong, was to take part in operations against Kikuyu terrorists in the 'Mau Mau' movement. Initially based at Muthaiga, they formed part of the 49th Infantry Brigade.

'Somewhat tired after a sweep above Naivasha': soldiers of the KSLI taking a brief rest after a patrol during 'Operation Bullrush' in January 1956. Anti-terrorist sweeps were made through the heavily forested hills of the Aberdare Mountains and around Lake Naivasha.

A Kikuyu girl admiring the tiger-skin apron of a KSLI drummer at Gatundu in Kenya. By November 1956, the KSLI was the only British Regiment still serving in Kenya. Apart from brief diversions to Bahrein and Aden, the Battalion remained there during the State of Emergency until June 1958, when it returned home.

Field Marshal Sir Gerald Templer inspects a Guard of Honour of 'C' Company, Muthaiga, in 1958. The men wear the linked 'K-K' flash of the 49th Infantry Brigade, denoting service in Korea and Kenya. Introduced in May 1955, the badge was first worn by the KSLI in June 1955.

Boarding for Bahrein in November 1956. While the Battalion was in Kenya, the Suez Crisis erupted and the KSLI was suddenly ordered to the Persian Gulf to deal with rioting on the island of Bahrein. This was quickly settled by the arrival of British troops and the Battalion returned to Kenya in January 1957.

Operations in Aden between April and May 1958. As the Battalion was preparing to leave Kenya, 'B' and 'C' Companies were called to Aden to help suppress a rebellion which had broken out near Dhala on the Yemen Frontier. The main action was 'C' Company's attack on the Jebel Jahaf on 30 April. The two Companies returned to England in June 1958.

The KSLI returning home: the troopship *Dunera* arriving at Southampton from Kenya on 26 June 1958. In the foreground is the band of the Somerset Light Infantry. The Battalion was greeted by the Lord Lieutenant of Shropshire, the Mayors of Shrewsbury and Hereford and former officers of the Regiment. The KSLI then went to Roman Way Camp, Colchester, as part of the 3rd Division.

Teaching the next generation. An aspiring bugler, Recruit Hayes, receives instruction on the trumpet from Bandmaster R. Ridings at the Depot. Bandmaster Ridings later became Lt Col. and Senior Director of Music of the Brigade of Guards.

Training continues. Up to 1960, National Servicemen were trained at Copthorne and other Light Infantry Depots such as Bodmin.

Two VCs on parade. Sgt George Eardley VC MM, formerly of the 4th Battalion, and Sgt Harold Whitfield, formerly of the 10th Battalion, taking the salute at Copthorne Barracks on 1 September 1956. Both men were frequent guests of honour at regimental functions.
Sgt Whitfield died later that year, on 19 December 1956, after a road accident. Eardley, after surviving a crash in 1964 when a train hit his car, died in Congleton in 1991. Standing to the right in this photograph is Maj. R.H. Garnett MBE.

After three years overseas, the KSLI returned to Shrewsbury in August 1958. Shown here is a representative detachment of the Battalion, marching up Pride Hill prior to a civic reception. The rest were stationed at Colchester. The next day, a combined parade with the Old Comrades was held in The Square.

Roles reversed on Christmas Day, 1958. At the Depot, Copthorne, an old tradition is being observed when the men are served drinks by the Depot Commanding Officer, Maj. R.H. Garnett MBE, and by the RSM.

The Counties March in Bridgnorth in 1960. In order to renew links within Shropshire and Herefordshire, the Battalion organized a march through the counties in April and May 1960. The routes through Shropshire took in Shrewsbury, Coalbrookdale, Madeley, Ironbridge and Wellington, followed by Oakengates, Dawley, Shifnal, then on to Broseley, Wenlock and Bridgnorth and finally to Oswestry and Ludlow.

New recruits in 1960. Note the new Light Infantry Brigade badge, a plain strung bugle, adopted by the KSLI and other county Light Infantry Regiments in place of their own distinctive badges in 1958. It remains the badge of the present Light Infantry.

Pay Parade at the Depot. A soldier receives his pay from Capt. K. Blockley, seated on the far left, watched by Lt Nicholas, seated next to him. Standing to attention is Sgt A. Stewart.

In Germany again. The KSLI returned to Germany in 1961 after nearly three years at home, and was stationed at Münster until 1964. Frequent training exercises at Company, Battalion, Brigade and even Divisional level, were carried out with other units of the British Army in Germany as well as with other NATO forces. Here NCOs enjoy some of the local bread. At the front, on the left, is Sgt Jim Griffiths and next to him 'Pete' Godbehere, who served from 1933 to 1968. Bandmaster Ridings is in frock coat, third from right.

Bugle Major B.J.T. Laidler, wearing the green No. 1 Dress uniform with white-metal badges adopted in 1947. This form of dress was based upon the special blue uniform worn at the coronation in 1937.

A Sign of the Times: even the Army could not escape the onslaught of pop music in the 1960s. The Junior Band of the Light Infantry Brigade at Copthorne, c. 1963.

Men of the 1st KSLI launching an assault craft on Lake Vogelsang, Germany, in September 1962, while on exercise with Belgian troops. National Service ended in 1962 and resulted in the loss of many valuable trained soldiers.

New Colours being presented to the 4th Battalion by Princess Mary, the Princess Royal, on 25 June 1964. Over 2,000 people witnessed the ceremony held in The Quarry. The Princess Royal had been present in 1911 when her father, King George V, presented the former Colours to the Territorial Battalion at Bangor.

The KSLI had already received the Freedom of Shrewsbury and Bridgnorth in 1955 and the Freedom of Oswestry and of Hereford in 1960. In May 1965, it was granted the Freedom of Much Wenlock, cementing the town's long-standing links with the KSLI. Here, General Musson, Colonel of the Regiment, can be seen handing the Freedom Casket to the Regiment. The casket contained an illuminated scroll recording the grant which is now held in the Regimental Museum. Note the Self Loading Rifles which came into service in 1955.

In the jungles of Malaysia in December 1966. The KSLI embarked upon what was to be its last overseas tour of duty in September 1966. Deprived of the opportunity of active service in Borneo, the Battalion went first to Singapore and then to Terendak near Malacca to carry out jungle warfare training. At Terendak, the Battalion once more served in the 28th Commonwealth Brigade and found itself alongside troops from Australia and New Zealand.

Two Malay boys keep an eye on Pte David Halford during exercises in Malaysia early in September 1967. Halford is seen with the General Purpose Machine Gun (GPMG) introduced in 1962.

Sgt Paddy Hannah with an Armalite rifle in Australia in 1967. The KSLI took part in Exercise 'Piping Shrike' in Queensland in September 1967. On this occasion, the men of the KSLI found themselves as 'the enemy' – in this case representing the Viet Cong!

A group visiting the Lone Pine Sanctuary near Brisbane in October 1967 – with a few friendly kangaroos. After Exercise 'Piping Shrike', the Battalion undertook goodwill visits to various Australian towns and always received a marvellous reception. The Band and Bugles were always well received.

The Battalion arriving at Enoggera station, Brisbane, on 10 October 1967, after completing exercises with the Australian army in the Shoalwater Bay training area.

Farewell to Australia: the KSLI marching through Brisbane on 12 October 1967. Taking the salute at the City Hall is Major General T.F. Cape, Australia's GOC Northern Command. The Battalion marched with fixed bayonets and with Colours flying, while the streets were packed with cheering spectators. It was a fitting end to the Australian tour.

The opening of the new Shirehall in Shrewsbury on 17 March 1967. With the Regular Battalion overseas, the Guard of Honour is composed of the 4th KSLI, commanded by Maj. S.E. Wardle (left). This was the last ceremonial function of the 4th (Territorial) Battalion which was disbanded only a fortnight later on 1 April.

An aerial view of Port St Louis, Mauritius. The last 'active service' duty which the KSLI undertook was in Mauritius in 1968. The Governor had declared a State of Emergency in January and called for British troops to help quell the rioting which had occurred prior to Independence. Two companies of the KSLI were rushed to the island from Singapore and quickly brought order to the streets of the capital. They remained on the island until after the Independence ceremonies on 12 March and most had returned to Malaysia by July.

A commemorative booklet for Sounding Retreat on 10 July 1968. This marked 'the final hours' of the KSLI. The Regiment had been fortunate not to be too deeply affected by successive armed forces reorganizations, but while it was in Malaysia the time came for its absorption into a new 'Large Regiment', The Light Infantry, as the 3rd Battalion. 'Vesting Day' for the new Light Infantry was celebrated with due ceremony at Copthorne on 10 July 1968 when the KSLI as such ceased to exist.

The silverware of the Officers' Mess on display in Malaysia in 1968. The items shown here cover almost 200 years of the Regiment's history; most are now on display in the Regimental Museum in Shrewsbury Castle.

The final parade. The 3rd Light Infantry marching through Bridgnorth on the 16 April 1970 *en route* to St Leonard's church to lay-up the Colours of the 1st KSLI. These were the Colours presented to the 1st Battalion at Göttingen in 1954. It was fitting that they should have been laid-up in St Leonard's as it was the parish church of the town in which the old 53rd had been raised in 1755.

The Band of the 3rd Light Infantry in Bridgnorth on 16 April 1970.

Old Comrades of the KSLI – 'that numerous, proud and indispensable body' – with Shropshire Cadets enjoying refreshments after the laying-up ceremonies.

St Leonard's church, Bridgnorth, 16 April 1970. In the commemorative booklet it said: 'The Regimental Call will be sounded. The Colours will be marched in slow time to a place before the Sanctuary while the Band plays the Regimental Slow March.'

The Colours, received by Maj. J.F. Hibbert of the 3rd Light Infantry, to be delivered to the Rector of St Leonard's 'for safe custody within this church'.

While the KSLI in name had gone, its former Depot at Copthorne, renamed Sir John Moore Barracks in 1963, continued to serve as the Light Infantry Depot. When the Depot was moved to Winchester in 1986, the Queen Mother, as Colonel in Chief of the Light Infantry, officiated at a full scale ceremony in which she reviewed the last soldiers to pass out of Copthorne Barracks. Her Majesty is seen here sighing the visitors' book.

The Queen Mother with Major General Lane, GOC South Western Command, at 'The Queen Mother's Gate' – a reminder of her visit in 1969. Once again known as Copthorne Barracks, the old KSLI Depot now fulfils an important role as HQ, Wales and Western District, and is also the home of the 5th (Shropshire and Herefordshire) Light Infantry (Volunteers).

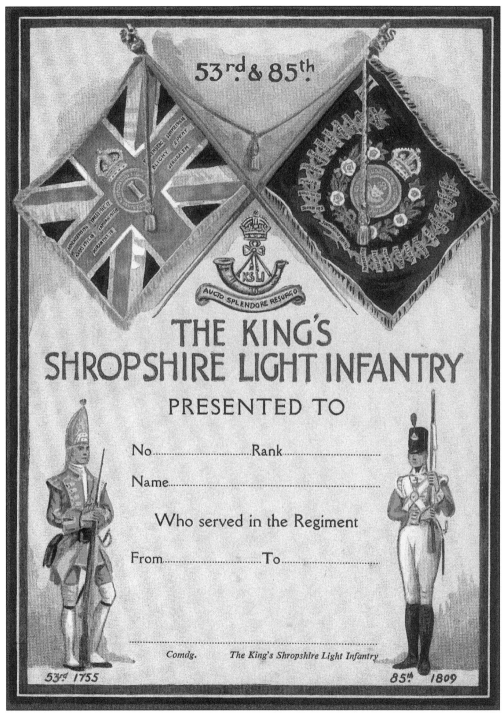

A Certificate of Service in the KSLI given to soldiers on completion of their service with the Regiment. It combines regimental images of many kinds: the Colours of the 1st and 2nd Battalions, the regimental badge, uniforms of former days and its forward-looking motto: 'Aucto Splendore Resurgo' – 'I rise again in increased splendour'.